Lara Letícia Galdino Amorim
Fabrício Almeida

Climate change and food security

Lara Letícia Galdino Amorim
Fabrício Almeida

Climate change and food security

A study applied to the milk production chain in Minas Gerais

ScienciaScripts

I dedicate this book first and foremost to God, for the strength, courage and maturity I have found in building this work.

To my husband Pedro, companion, encourager and friend at all times.

To my parents Romildo and Ivete and my brother Fernando, who are the moral and emotional foundation of an exceptional family.

CONTENTS

CHAPTER 1
INTRODUCTION

The growth of the world's population has led to a number of changes in production and international trade, especially the expansion of the food sector. In fact, this issue has fuelled an important debate on food and nutrition security (FNS) in several countries, justifying extensive investment in agriculture, livestock, infrastructure and the processing and distribution industries.

In order to meet global food and nutrition security, food production needs to grow substantially. However, even with a large increase in production, many people around the world do not have access to food, due to poverty, low purchasing power and the high price of industrialised food.

Milk is a food that deserves attention when it comes to food safety because, among other factors, the nutritional value of this product and its derivatives is essential for meeting some of the population's food needs, since they are sources of energy, proteins, carbohydrates, fats, vitamins and minerals.

In Brazil, one alternative for guaranteeing food security would be to jointly study population growth and the volume of food available to meet the population's needs. However, these studies alone are not enough to guarantee this, since there are other factors that can limit the quality and quantity of some foods. These limitations are due to technological, socio-economic, institutional and edaphoclimatic factors, which have been changing as a result of climate change.

Over the years, climatic variations have been studied with greater emphasis, as the importance of climate in various activities has been realised, such as agro-industrial activities, since variations in temperature or rainfall can have a significant effect on industries' choice of a particular input, and on farmers' preference for certain crops and livestock that may suffer less from climatic variations.

Therefore, among other factors, the discussion about policies and strategies aimed at food security already includes studies and research into the effects of climate change on the production sector. In the case of the agricultural sector, this debate is of paramount importance, considering the capacity of climate variables to intervene in production, crop yields and livestock breeding.

This research therefore aims to assess the implications of regional climate change for the milk production chain in the state of Minas Gerais.

1.1   OBJECTIVES

1.1.1 General Objective

The general aim of this work is to analyse the possible impacts of climate variables for the state of Minas Gerais on the milk production chain.

1.1.   2Specific Objectives

The specific objectives of the work are:

-Organise and evaluate the climatological data and historical series of interest and available from January 2005 to December 2014;

-Propose   statistical modelling capable of demonstrating the transmission of climatic variations along the production chain;

Discuss the   possible implications of climate change for food security in a short- and medium-term scenario.

1.2   BACKGROUND

It is believed that carrying out this research is important because observing climate change presupposes that it could affect food production. With this in mind, we question the capacity of climate aspects to intervene in food production, especially in the milk production chain.

Similarly, can the rainfall measured in a given region influence the quality and volume of feed and inputs available to the herd? Can variations in temperature influence the behaviour of animals and in some way alter their zootechnical performance?

Given these questions that guide this research, the interest in climate studies and their impact on the ability to ensure reliable production levels in line with economic, social and environmental demands is justified.

1.3   THEORETICAL BACKGROUND

1.3.1   Food and Nutrition Security

The conceptual development of food security takes place as a continuous process, in permanent construction, varying with the disparate needs of each nation and evolving throughout history.

It is possible to say that Thomas Robert Malthus was one of the first thinkers to address, albeit indirectly, the issues of food security in connection with population growth, since Mathus published the Essay on Population in 1798, and his theory was based, among other arguments, on the idea that food production would grow in arithmetic progression while population growth would obey an uninterrupted geometric progression. Based on this theory, he "predicted" that the volume of food would not be enough to satisfy the population's food needs (ALENCAR, 2001).

However, it was in Europe that the concept of food security began to be used during the First World War (1914 - 1918), which associated this term with national security and the productive capacity of each country (ABRANDH, 2010).

However, this concept gained momentum after the Second World War (1939-1945), especially after the constitutionalisation of the United Nations Organisation (UNO). It was at this time that food insecurity caused by the scarcity of food production in poor countries began to be assessed (ABRANDH, 2010).

The global food production crisis of the 1970s prompted the 1974 World Food Conference to evaluate food production and its regularity of supply, as there is a difference between availability and access to food, as it may be available, but some populations, especially those on lower incomes, may not have access to it (ABRANDH, 2010).

In the 1980s, the concept took a new direction, as [...] food security came to be related to guaranteeing everyone physical and economic access to sufficient quantities of food (ABRANDH, 2010, p. 12).

According to the declarations made by the Food and Agriculture Organisation of the United Nations (FAO) and the World Health Organisation (WHO) at the 1st International Conference on Nutrition in 1992, the concept of food security began to be interlinked with other terms such as access to safe food (free from contamination), of quality and produced in sustainable ways, as well as incorporating the idea of access to information, thus adding nutritional and health aspects to the concept, which came to be called Food and Nutrition Security (VALENTE, 1997).

Recently, the concept of food sovereignty has been associated with the term FNS. Food sovereignty defends the right of each country to propose efficient policies that guarantee the Food and Nutritional Security of its population, including the right to preserve the traditional production practices of each culture (ABRANDH, 2010).

In Brazil, the term Food and Nutrition Security began to be publicised with greater emphasis at the end of the 1990s, due to the country's preparations for the World

Food Summit and the creation of the Brazilian Food and Nutrition Security Forum (FBSAN).

Currently, the concept of FNS is established by the aforementioned Article 3

of Law 11.346/2006 (Organic Law on Food Security - LOSAN):

Food and nutrition security consists of the realisation of everyone's right to regular and permanent access to quality food, in sufficient quantity, without compromising access to other essential needs, based on health-promoting eating practices that respect cultural diversity and are environmentally, culturally, economically and socially sustainable (BRASIL, 2006).

The National Food and Nutritional Security System (SISAN) in Brazil was set up following the creation of this law, which led to a wide-ranging discussion on FNS aimed at creating a SISAN that would guarantee the desired food and nutritional security for all Brazilians (MACEDO, et al., 2009).

1.3.2 Dairy Production and Food and Nutrition Security

Milk and its derivatives should be part of everyone's diet, as it is an ideal source of nutrients that can help to fulfil some of the population's dietary needs while also helping to guarantee food security.

From a biological point of view, milk can be considered one of the most complete foods, as it is a source of essential elements such as micronutrients, amino acids, minerals, vitamins, proteins and a high calcium potential.

According to the Regulations for the Industrial and Sanitary Inspection of Products of Animal Origin - RIISPOA, article 475, "milk, without any other specifications, is understood to be the product derived from the complete, uninterrupted milking, under hygienic conditions, of healthy, well-fed and rested cows" (BRASIL, 1952).

Drinking milk at all stages of life contributes to physical and intellectual development, because according to nutritionist Sandra Almeida (2011), dairy products are important for the growth of bones and teeth in children. For adults, this food helps prevent the loss of bone mass (osteoporosis).

Milk protein, casein, is considered a complete protein of excellent quality and a source of essential amino acids for the growth and maintenance of the body's tissues (SINDILEITEPR, 2013).

Of the minerals found in milk, calcium and phosphorus contribute to the structure of bones and teeth. Fats contain fatty acids, which help with the absorption of fat-soluble vitamins (SINDILEITEPR, 2013).

Among the vitamins available in milk in greatest quantity are vitamins A, which protects the skin and eyes, helps to form body tissues and keeps hair healthy, and some of the B complex, which are important for energy production, oxygenating cells and protecting the nervous system (SINDILEITEPR, 2013).

So the nutritional factor, among other characteristics, is the most important component for this food and its derivatives to be among the foods that can provide food and nutritional security.

In this context, it is important to make a critical analysis between the largest milk-producing regions and the regions that have the greatest demand for this food, as a divergence in these can have a negative impact on food and nutritional security.

In the case of Brazil, the regions that produce the largest volume of milk are the southeast, centre-west and south, but it is the northeast that would require the largest volume of this food in order to eradicate hunger, because according to IPEA (2013) this would be the region with the greatest poverty in the country, so, among other factors, the logistics of this input from one region to another can cause depreciation, increase in price and decrease in the quality of the product, thus having a negative impact on food and nutritional security.

In Brazil, the Incentive Programme for Milk Production and Consumption (PAA) was created in 2003 as an initiative to combat hunger. The programme aims to combat malnutrition among children and pregnant women. The PAA is an example of the importance of milk in the sphere of food security.

1.3.3 Organisation of   the Milk Production Chain in Brazil

Dairy farming began to gain national prominence in the mid-60s. But it was around 1980 that the sector exhibited

a dynamism that allowed the chain to progress even with the direct intervention of the Federal Government in the production of the product.

Until the end of the 1980s, the federal government intervened directly in milk production, regulating prices and controlling imports (FIQUEIRA, BELIK, 1999).

From the 1990s onwards, the dairy industry began a process of intense transformation, resulting in important actions to strengthen its production chain, such as the introduction of integrated logistics and the restructuring of the chain's links. The growth in production was the result of competitiveness in an open market with free prices, unlike in the past when the increase in production was the result of an increase in the herd (CILEITE, 2015).

According to Figueira and Belik (1999), the 1990s were marked by changes in the institutional environment of the milk production chain, due to the state's new stance on the economy, which led, among other transformations, to the liberalisation of prices at all links in the chain, the expansion of the market and the implementation of MERCOSUR (Southern Cone Common Market).

It is therefore clear that in recent decades dairy farming relations have undergone changes. Nowadays, where economic globalisation is the driving force behind agro-industrial activities, market flows have to be analysed with greater emphasis, in a complex system of production chains (EMPRABA, 2005).

In Brazil, especially in the milk production chain, there are certain production limitations and bottlenecks that can somehow interfere with the food security model since the country is among the five largest milk producers in the world (IBGE, 2013).

The milk production chain in Brazil involves various sectors, as shown in

Figure 1, including family farming made up of small producers, large producers, the production of inputs (soya and maize) to feed herds and the marketing and industrialisation of the product and its derivatives.

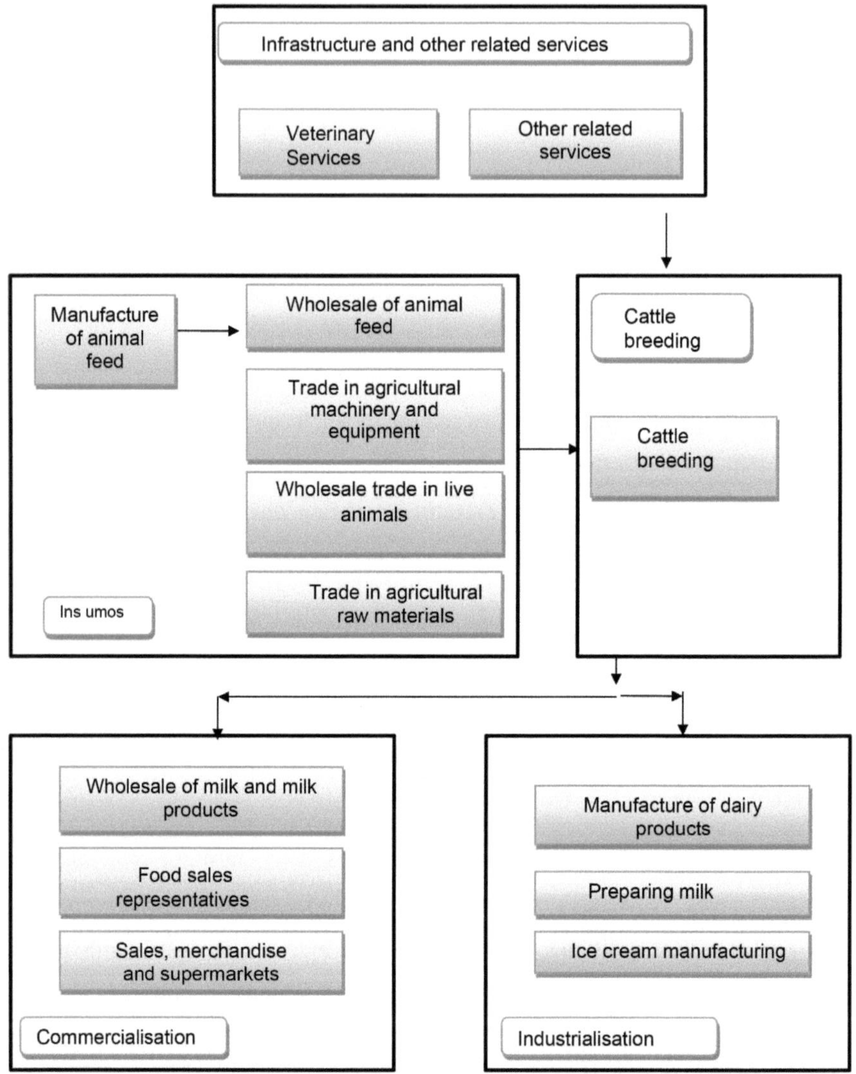

Figure 1- Brazil: Composition and Structure of the Milk Agro-Industrial System.
Source: Adapted from Almeida (2007); SEBRAE (2007).

The links involved in the production chain must prioritise certain important points in order to compete on an inter- and extra-sectoral level. These include training, professionalism and administrative and managerial competence (EMBRAPA, 2005).

Faced with this agribusiness scenario, knowledge of the milk Agro-Industrial System (SAG) flows is essential for guaranteeing the market and commercialisation of the product, as it is by analysing the segments of the chain that it is possible to identify the limitations of the activity, assess input prices, find new market niches and define more competitive prices (EMBRAPA, 2015).

It is therefore extremely important that there is harmony between the links in the milk production chain, as disorganised actions within the chain can hinder product quality control, the progress of production systems, the growth and structuring of the dairy basin and the generation of income and services (SIMÃO NETO et al., 1989; TOURRANT et al., 1998; VEIGA et al., 2001).

According to Almeida (2007, p. 28), some factors are important for understanding and sizing production when considering the Milk Agro-Industrial System: "[...] the profile of the production units; the role of the agro-industrial-dairy industry in the acquisition, processing and commercialisation of by-products; adding value in the production chain". It's worth stressing that a SAG is based on the idea of systemic organisation and coordination of the agri-food production chain.

Dairy farming in Brazil is characterised by a distinct feature: there is no production standard, i.e. national milk production is divided between large and small farmers.

Another important factor for the milk SAG is seasonality, since the country has a predominantly tropical climate, which is characterised by high temperatures and well-defined seasons, such as dry winters and rainy summers (MOREIRA, 2002).

The hot, rainy summer is characterised by an abundance of forage in the pastures, directly influencing the volume of milk, since this is the main food for the herds. Dry winters with a lack of rain and cold weather can be the main cause of the drop in milk volume in the off-season, due to the reduced availability and nutritional quality of pastures, resulting in the animals being fed supplementary concentrates (JUNQUEIRA et al., 2008).

The seasonality of milk production can have implications for the milk agro-industrial system by reducing producers' income due to the drop in milk volume in the dry season, while at the same time it can increase production costs due to the need to provide cattle with supplementary concentrates and higher labour costs.

The dairy sector is experiencing a period of expansion in its production, attributed mainly to the production efficiency conferred by more competitive profiles and higher quality, at the same time as market technologies and innovations are incorporated into the dairy chain (ALMEIDA, 2007).

In addition, the dairy industry has been undergoing reorganisation processes in order to make the production chain more efficient, innovative and competitive in the global market. As a result, new concepts have emerged about productivity, costs and agility in order to ensure that we remain in increasingly competitive and globalised environments.

1.3.   4General Aspects of   the Dairy Basin in the State of Minas Gerais

Milk production in Brazil is distributed throughout the country, but the largest dairy basins are concentrated in the south, southeast and centre-west regions.

It is possible to say that Minas Gerais is a privileged state when it comes to milk production, since the state has a historical and political vocation to develop this activity. An example of this is the Coffee with Milk Policy during the Old Republic - the agreement was named after the economies of Minas Gerais and São Paulo, major producers of milk and coffee respectively.

In addition to this, Minas Gerais has actions to strengthen dairy farming, such as the State Milk Production Chain Programme, better known as Minas Leite. The programme is aimed at family farmers who have dairy farming as their economic base.

The programme is implemented through the Minas Gerais State Technical Assistance and Rural Extension Company (Emater-MG), which guides producers to improve productivity by using low-cost technologies.

Another action that contributes to the dairy sector is the Programme to Improve the Genetic Quality of the Bovine Herd in the State of Minas Gerais (Pró-Genética). Created by the Secretary of Agriculture, Livestock and Supply (Seapa), the programme aims to strengthen the meat and milk production chains by introducing bulls with proven genetics and guaranteed by the Brazilian Association of Zebu Breeders (ABCZ) onto farms.

According to the Brazilian Institute of Geography and Statistics (IBGE) (2013), the south-eastern region showed an increase in its share of the cattle herd in 2011 and 2012, while in 2013 the region's share of the cattle herd stabilised at 18.6% (Table 1), and it is worth noting that the largest increases were recorded in the state of Minas Gerais (11.4%).

Table 1 - Share of cattle - 2010 to 2013

Region	Share of cattle population (%)			
	2010	2011	2012	2013
South East	18,3	18,5	18,6	18,6

Source: Adapted, IBGE, Research Directorate, Agricultural Coordination, Municipal Livestock Surveys, 2010-2013.

According to IBGE (2013), of the total number of cattle in Brazil (211.764 million head), 10.8 per cent corresponded to cows that were sorted in 2013, with the Southeast region having the highest percentage (20.6 per cent) of the total number of cattle in the region, i.e. of the 18.6 per cent share of cattle in the country, the Southeast region has 20.6 per cent dairy cows, which demonstrates the region's dairy vocation.

Brazil saw a 6.0 per cent increase in milk production between 2012 and 2013. The increase affected all geographical regions, with the South and Southeast contributing the most to this growth. The regional share of milk production, as shown in Graph 1, shows that the Southeast produced the largest volume of milk in 2013 compared to the other regions of the country (IBGE, 2013).

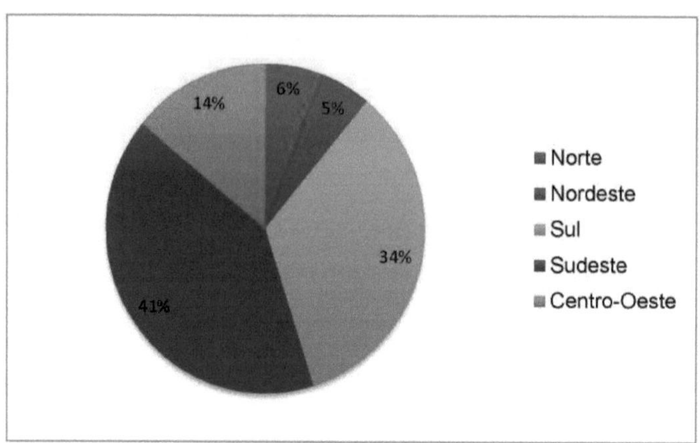

Graph 1 - Regional share of milk production in 2013

Source: Based on data provided by the IBGE, 2013.

The Southeast region accounted for 35.1 per cent of the amount of milk produced in Brazil in 2013, and the state of Minas Gerais was responsible for 27.2 per cent of this production, as shown in Graph 2 (IBGE, 2013).

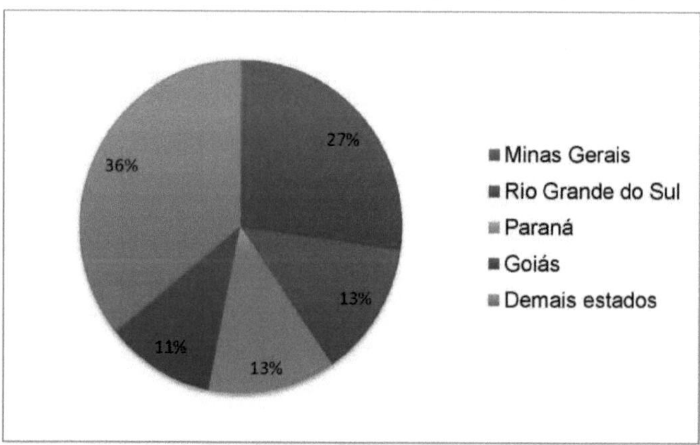

Graph 2 - State share of milk production.
Source: Based on data provided by the IBGE, 2013.

Among the states that have made a significant contribution to the

national milk production in 2013 are Minas Gerais, Rio Grande do Sul, Paraná and Goiás. Two southern states and one southeastern state account for more than 50 per cent of the total national production of this input.

The purchase of domestic milk by milk processing industries in the second quarter of 2014, compared to the same period in 2013, shows an increase of 8.4 per cent. The Northeast, Southeast and Centre-West regions were the ones that contributed to this increase, since the North and South regions showed a reduction in the share of milk purchases, as shown in Table 2 (IBGE, 2014).

Table 2 - Share of milk purchases - 2nd quarters of 2013 and 2014

Regions	2nd quarter 2013	2nd quarter 2014
North	5,8	5,1
North-East	5,1	5,7
South	34,2	33,8
South East	40,9	41
Centre-West	14	14,4

Source: Adapted, IBGE Indicators, Livestock Production Statistics, 2014.

The data shown in Table 2 indicates that the Southeast and South are the regions that purchased the most milk in the period analysed, with the Southeast accounting for more than 40% of national milk purchases.

Among the states that make up the Southeast region, Minas Gerais was the one that stood out in terms of its contribution to the milk acquisition indicator in Brazil. The state accounted for almost 90 per cent of the regional increase, which in a comparison of states would represent around 30 per cent (Graph 3) of the national total (IBGE, 2014).

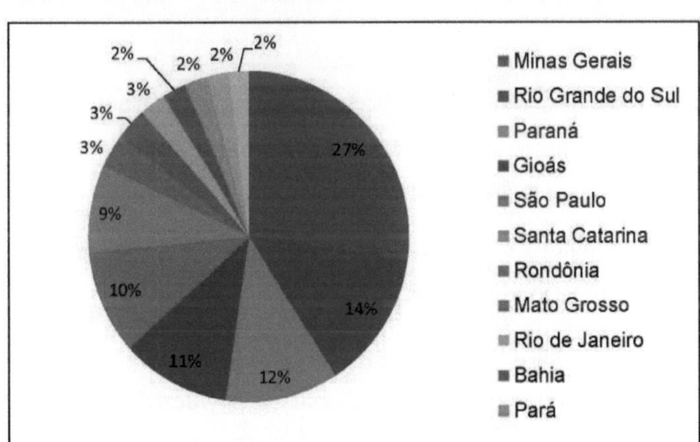

Graph 3 - Ranking of milk purchases - 2nd quarter 2014
Source: Based on data provided by the IBGE, 2014.

The state of Minas Gerais leads the way in terms of milk purchases, followed by Rio Grande do Sul, Paraná, Goiás, São Paulo, Santa Catarina and other states that don't have a significant volume of purchases.

Analyses of the data show that the state of Minas Gerais is among the country's largest dairy basins and can also be considered the state with the highest production of this input, as well as being the state that buys the most of the product.

The other states that make a significant contribution to national milk production are also the ones that buy the product the most, most often for industrialisation and later export.

Of the total volume of milk produced in Brazil, 68.8% was industrialised in 2013, with an increase of 8.2% in the second quarter of 2014 (IBGE 2013; IBGE, 2014).

This production and procurement scenario basically revolves around the same regions (Southeast, South and Centre-West), and the high rate of industrialised and exported products can be worrying from a food security perspective.

The Northeast region should be ranked among the regions that buy the most of this food, as it has one of the highest rates of poverty in the country, resulting in malnourished people who need a balanced diet, and milk, through its nutrients, could strengthen this diet.

Exports of dairy products in Brazil have been increasing significantly in the country and could double with the new Healthy Milk Programme launched by the Federal Government in September 2015. The programme's proposal, among other actions, is to help increase the quality and volume of milk produced in order to boost exports of dairy products by 2018 (PORTAL BRASIL, 2015).

This action could have an impact on guaranteeing food security, since industries will prioritise the production of export foods such as powdered milk, butter and cheese instead of producing foods aimed at a balanced diet for Brazilian society.

1.3.   5Global Climate Change   and Interventions in Economic Activities

For some years now, there has been intense concern about greenhouse gas (GHG) emissions due to

mainly due to the phenomenon that these emissions are associated with, climate change.

According to the Intergovernmental Panel on Climate Change (IPPC) (2007) the term climate change refers to changes in the state of the climate that can be identified by average changes and/or in the variability of its properties, and that persist over an extended period, typically decades or more.

GHG emissions may be associated with deforestation and the opening up of new agricultural areas, the practice of incorporating organic crop residues and turning over the soil for cultivation and, finally, extensive animal husbandry combined with agricultural activity.

In this respect, the phenomenon of GHG accumulation in the atmosphere, among many other factors, can be considered, from an environmental and economic point of view, the most relevant for understanding the effects of climate change on the environment on a local scale. Therefore, the profile of GHG emissions by country or economic bloc must be considered, taking into account current environmental policies, economic activities, as well as the technological standard and other resources allocated to productive arrangements.

According to the IPPC (2007), changes in the amount of GHGs and aerosols in the atmosphere, as well as variations in the emission of sunlight on the earth's surface, interfere with the energy balance of the climate system.

As indicated in the Specifications of the Brazilian GHG Protocol Programme, the internationally recognised greenhouse gases regulated by the Kyoto Protocol are:

> Carbon dioxide (CO_2);

> Methane (CH_4);

> Nitrous oxide (N_2O);

> Sulphur hexafluoride (SF_6);

> Hydrofluorocarbons (HFCs);

> Perfluorocarbons (PFCs).

Global concentrations of carbon dioxide, methane and nitrous oxide have increased considerably as a result of anthropogenic activities. This increase in CO_2 concentration is mainly due to the burning of fossil fuels and changes in land use, while the increase in CH4 concentrations is mainly due to agriculture (IPPC, 2007).

In the case of Brazil, the emissions scenario is different from most other countries, because according to the Amazon Research Institute (IPAM) (2015) the amount of emissions in Brazil from burning fossil fuels is relatively low when compared to the amount emitted by other countries. This is because Brazil's energy matrix is considered relatively clean by international standards, since it is based on hydroelectricity. However, most of Brazil's emissions are the result of land use activities, such as deforestation and burning, which most often occurs for agricultural expansion.

With this in mind, the Report of the Technical Panel of the Ministry of the Environment on financing, benefits and co-benefits (2012), considers forest degradation and deforestation to be one of the world's main sources of greenhouse gases.

Climate change, especially global warming, is caused by the expansion of carbon gases, methane and nitrous oxide in the atmosphere (IPCC, 2007). The climate can act as a regulator of production efficiency in the agricultural sector and can have negative effects, especially on dairy farming.

In 2013, the agricultural sector felt the effects of climate change. One example of these effects was the intense drought that occurred in historically producing regions, which partly reduced the supply of animals and milk production, and consequently influenced the production of its derivatives (IBGE, 2013).

The study of climate change has therefore received special attention at international conferences and political discussions, with the aim of reducing the vulnerabilities of agro-industrial systems of interest.

Changes in the production environment, quality standards and food safety require the implementation of strategies to assess the possible effects of climate change.

Changes in climate can affect dairy farming through higher prices and reduced availability of grains, the mobility of pests and diseases that reduce the quality and quantity of forage plants, which are the basis of cattle feed, as well as intensifying heat stress in animals, resulting in reductions in reproduction, gestation, food consumption and the productive efficiency of dairy cows (SILVA et al., 2009).

It is worth noting that the animals' response to a stressful event is based on three main elements: "recognising the threat to homeostasis or well-being, responding to the stress and the consequences of the stress" (EMBRAPA, 2009, p. 17).

Climate stress, according to Baccari Junior (1998), is stress caused by climatic factors, which can cause changes in growth, milk production and reproduction in herds. For example, in hot climates, heat stress prevails, i.e. the thermal stress generated by high temperatures and high relative humidity in conjunction with high metabolic heat production, resulting in surplus body heat and, if it is not possible to eliminate this surplus, heat stress occurs (EMBRAPA, 2009).

As Pires et al. (1998) point out, when dairy cows are exposed to heat stress, food consumption and basal and energy metabolism are inhibited, while body temperature, the respiratory process and sweating rate increase. These changes are indicators that the animal is trying to reduce the thermal imbalance in order to maintain homeothermy, and the duration and intensity of the stressor determine the animal's response (behavioural, physiological and immunological) to the stress (EMBRAPA, 2009).

According to the Brazilian Agricultural Research Corporation (2009, p.20) "physiological responses to heat include peripheral vasodilation, an increase in the rate of sweat production (sweating rate), an increase in respiratory rate and a reduction in energy metabolism".

In the immunological field, the combination of temperature and humidity causes diseases that make the immune system more vulnerable, contributing to the emergence of diseases and their spread (REIS, 1996).

Behavioural responses to heat stress can be seen when herds reduce their activities during the hottest periods, preferring shaded areas or immersion in water, as well as choosing to graze at night due to the more pleasant climate (EMBRAPA, 2009).

According to EMBRAPA (2009, p.20), the effects of heat stress on dairy cattle:

> Their activities are compromised, which leads to changes in food and water consumption, growth/development, milk production and reproduction, as well as affecting their behaviour (physical activity, body posture, seeking shade). Higher production cows are more susceptible to heat stress due to the greater heat production resulting from higher food intake to meet the high production demand.

On the other hand, according to Hahn (1993), cattle's responses to climatic stress caused by cold are basically to prevent heat loss and increase heat production.

CHAPTER 2

METHODOLOGY

This research was carried out using quantitative analyses on historical series of interest, with the aim of correlating the implications of climate change for milk production. In addition, the transmission of these probable impacts on the price paid to milk producers and on the price of its derivatives was analysed.

The econometric time series model used consolidated monthly data from January 2005 to December 2014. The series are shown in Table 1, along with their description, source of recording and unit of the values found.

Table 1 - Description of the variables used in the econometric model.

Variables	Series Description	Source	Unit
PPTM	Total monthly rainfall	BDMEP	mm
TEMP	Average monthly temperature	BDMEP	⁰C
PPPL	Average nominal price paid to milk producers	CEPEA	R$/thousand l
PMDA	Average wholesale price of dairy products - whole raw milk	CEPEA	R$/thousand l
PRODUCTION	Monthly milk production	CEPEA	mi l

Source: Prepared by the authors, 2015.

The series described refer to the average nominal price paid to milk producers, the average wholesale price of dairy products and monthly milk production. We chose to analyse whole raw milk as a by-product because of its nutritional value for people in terms of food security. The climatic factors considered in the analysis relate to average monthly temperature and total monthly rainfall.

The data was subjected to the unit root test (Augmented Dickey-Fuller Test - ADF) in order to verify the stationarity of the time series using an auto-regressive model. Since the VAR model (auto regressive vectors) for time series studies requires the data to be stationary (when the mean and variance of the information become constant over a longer period of time and the value of the covariance between the time of the two periods depends on the distance or difference between the two periods and not on the actual time at which the covariance is calculated) (GUAJARATI, 2000).

The ADF test is defined by the following equation (Dickey-Fuller, 1979):

$$\Delta Y_t = \beta_1 + \beta_2 t + \delta Y_{t-1} + \alpha \sum_{t-1}^{p-1} \Delta Y_{t-1} + \varepsilon_t \qquad (1)$$

where β_1 is the intercept, t is the trend; Δ is the difference operator. The lag order (p) is found using the Schwarz Criterion (SC), in order to obtain uncorrelated residuals, i.e. white noise (ENDERS, 1995). According to the same author, the Schwarz criterion is the most parsimonious.

In models with seasonality, the most direct method is when the seasonal pattern is purely deterministic (ENDERS, 1995). Thus, S1, S2 and S3 are considered to represent the dummies for each semester, in which the value of Si is equal to one in period i and zero in the other periods. The estimated regression is represented by:

$$\Delta Y_t = \beta_1 + \beta_2 t + \delta Y_{t-1} + \alpha \sum_{t-1}^{p-1} \Delta Y_{t-1} + S_1 + S_2 + S_3 + \varepsilon_t \qquad (2)$$

Seasonal dummies for milk were defined for this study as S1 (November to February), S2 (March, April, September and October) and S3 (May to August), due to the similarity of the periods in terms of rainfall.

Subsequently, lag estimation was applied to the model, using Schwarz's Bayesian Criterion (BIC) as a reference, which weights the maximised log-likelihood function and the number of model parameters (SCHWARZ, 1978). The best model is the one with the highest BIC value given by:

$$BIC = E[\ln L(\theta)] - \frac{1}{2} p \ln(N) \qquad (3)$$

where E [In L(θ)] is the expected value taken with respect to the a posteriori density of the log-likelihood function, p is the dimension of the parameter vector and N is the sample size or data series.

Considering the interdependence between the time series studied, it is necessary to estimate a VAR model. A vector auto regression model is a set of k time series regressions where the regressors are the lagged values of all k series. In this way, it can be said that the future value of a variable is predicted from the lagged values of itself or other variables in the system. Two or more variables are used to make these predictions of future values.

Some considerations about the model revealed that since VAR is an atheoretical proposal, it should allow for the use of a robust set of data and with an analytical bias in various senses, allowing for a certain coherence in the interpretation and evidence in the forecasts obtained from the general model. In this way, it can be said that the great practical challenge of this modelling is deciding on the appropriate lag length, in addition to the requirement that all the variables analysed are concomitantly stationary (GUJARATI, 2000).

The methodology used determines the response of the dependent variable of the VAR system to shocks in the stochastic error terms ε_{1t} and ε_{2t}. The transmission of shocks can have a greater effect in the future than in the initial periods on the original endogenous variable, due to the retroactive effects through other variables. With this, the effect that changes in innovations have on the present and future values of endogenous variables can be identified, determining the impact of such shocks over several predictive periods.

Thus, the decomposition of the forecast error variance for each variable is only relevant when determining the influence of each shock on the set of VAR variables.

In order for the variable chosen to be both dependent and independent in an empirical model, Granger (1969, apud ENDERS, 1995) suggests a concept of causality where the variable xt helps to predict the variable yt, aiding estimation. Thus, yt is called caused, in the Granger sense, by xt. The method consists of estimating the variables in the dependent variable position, one at a time.

$$Y_t = \gamma + \sum_{i=0}^{n} \alpha_i X_{t-1} + \sum_{j=1}^{m} \beta_j Y_{t-j} + \varepsilon_t \qquad (4)$$

$$X_t = \theta + \sum_{i=0}^{p} \rho_i Y_{t-1} + \sum_{j=1}^{q} \lambda_j X_{t-j} + \varepsilon_{2t} \qquad (5)$$

where Yt, Xt are the prices of interest; γ, α, β, θ, ρ, λ are the parameters to be estimated; i and j are the number of price lags; and et are the uncorrelated random errors.

If variable X Granger-causes Y, then changes in X must precede changes in Y. There are four cases of causality: (i) unidirectional causality from X to Y, (ii) unidirectional causality from Y to X, (iii) bilateral causality and, (iv) independence.

Finally, the VAR model was estimated by defining the impulse response functions.

The decompositions of the series of interest in the auto regressive model (VAR) for the state of Minas Gerais, as well as the transmission and prediction graphs, were generated by the GRETL programme (GNU REGRESSION, ECONOMETRIC SAND TIME-SERIES LIBRARY, 1.10.1, 2015).

CHAPTER 3

RESULTS AND DISCUSSIONS

3.1 Behaviour of Seasonality in Historical Series

In order to verify the possible interference of seasonality in the milk production chain, the decomposition of seasonality was observed in the historical series of milk production and total monthly rainfall.

Graph 4 shows the decomposition of seasonality for the historical series of milk production in the state of Minas Gerais from 2005 to 2014.

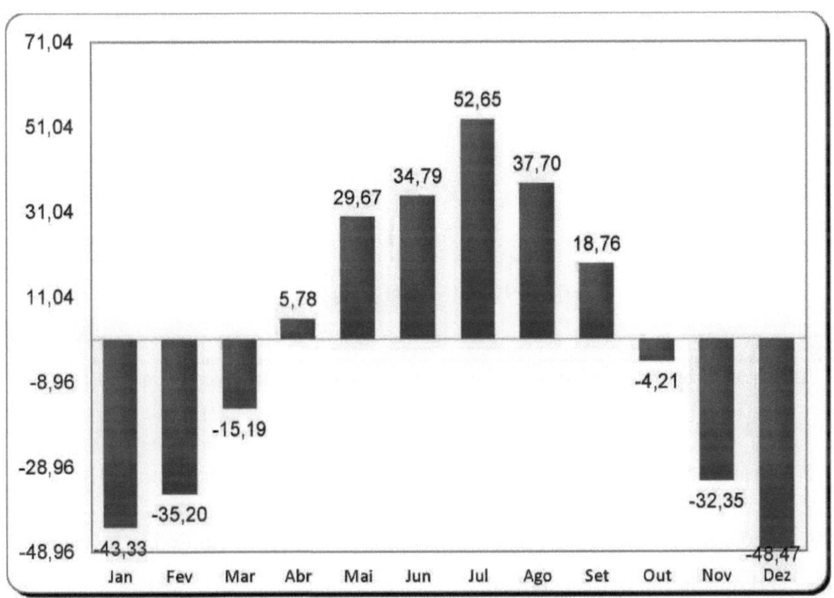

Graph 4- Decomposition of seasonality for historical series of milk production.

Source: Prepared by the authors, 2015.

The graph shows the influence of seasonality on milk production, determining distinct periods of higher (positive) and lower (negative) levels. The period from April to September shows a greater supply of agricultural products, even though it is the dry season. This is because the milk production response, and consequently the greater supply and availability of the product in the market, is due to the months prior to the greater rainfall.

In order to better understand the possible interference of seasonality in the historical series, it is extremely important to analyse Graph 5, which shows the decomposition of seasonality for the historical series of total monthly rainfall.

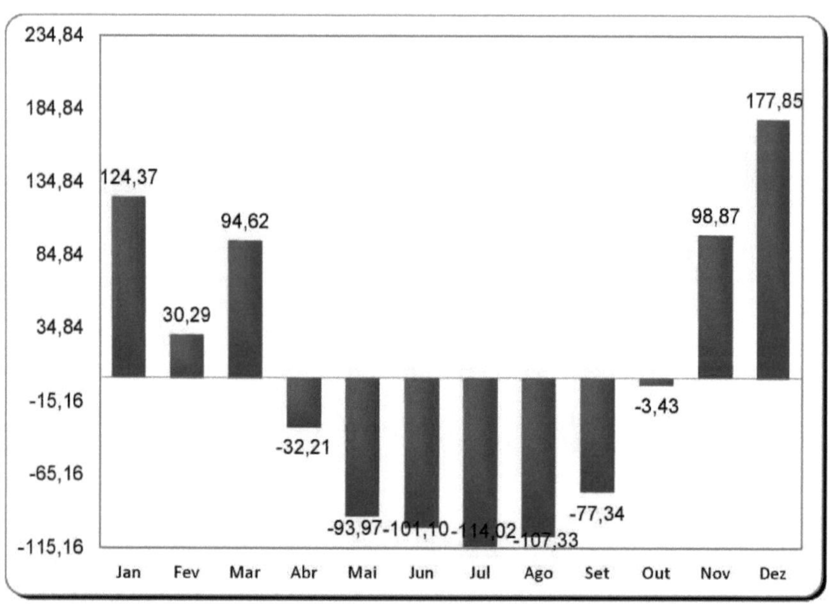

Graph 5- Decomposition of seasonality for the historical series of total monthly rainfall.

Source: Prepared by the authors, 2015.

The graph shows the influence of seasonality on the historical rainfall series.

It is possible to say that the seasonality of the climate is directly related to the rainfall rates achieved during the period studied.

Analysing the two graphs together implies that the period from June to August sees the greatest productive responses from the Minas Gerais herd, considering the accumulated rainfall in the period leading up to the drought. This residual effect, i.e. an increase in production in subsequent months, demonstrates the sector's dependence on climatic factors (average temperature and rainfall) and the slow and gradual response of production capacity in relation to the period of greatest rainfall concentration.

3.2 Verification of Stationarity

In order to check for stationarity, the Augmented Dickey-Fuller (ADF) test was carried out, the results of which are shown in Table 3, as well as the stationarity parameters for the variables in the historical series.

Table 3 - Summary of the Augmented Dickey-Fuller (ADF) unit root tests for the econometric model series.

Variables	Lag (p-1)	ADF - T Statistic		Stationarity (p< 0.05)
PPTM	1	T	-3,08606	Stationary
TEMP	1	T_μ	-6,30535	Stationary
PPPL	1	T_t	-5,24773	Stationary
PMDA	1	T_t	-4,23137	Stationary
PRODUCTION	1	T_t	-6,48162	Stationary

Source: Adapted from GRETL, 2015.

It can be seen that all the test results are significant, regardless of the type of market (at the levels of the climate series, the average nominal price paid to milk producers, the average wholesale price of dairy products and monthly milk production), i.e. the historical series selected to make up the VAR model are stationary.

The lags for the auto regressive model (VAR) were selected using the Bayesian Schwarz criterion (BIC), as shown in Table 4.

Table 4 - Selection of lags using the Bayesian Schwarz criterion (BIC) for the set of series of interest in the model.

Gaps	Log.L	p (LR)	BIC
1	-2359,05622		43,953841[1]
2	-2303,43860	0,00000	42,020671

[1] indicates the best (i.e. minimum) value of the information criterion.

Source: Adapted from GRETL, 2015.

The auto-regressive vector models will be described considering a 1-month lag, without constant, with constant and constant and trend, thus composing a regression model, where it is hoped, at least empirically, that it will be able to explain the composition of influence of the general model of variables.

3.3 Estimation of the VAR Model

The estimates of the VAR model consider the dependent variable to be the one in which the impact is to be verified, as a function of the shock hit in the set of series, thus allowing the transmission of this shock in the model to be assessed, understood as a factor causing the impact. The data outputs of the analysis and the results of the significant variables are described in Table 5.

It is worth noting that all the results were analysed in relation to the regions of the state considered in the research and between the temperature (18°C to 26°C) and rainfall (0mm to 450mm) ranges that occurred during the periods studied.

An empirical interpretation of the model, described in Table 5, suggests that the significant variables have an effect on the dependent variable. The fact that there is a one-month lag means that the model is able to transmit and perceive the effects. It is worth emphasising that the table above only shows the variables that have some significance when exposed to a shock.

Table 5 - Estimates of the VAR model for the set of climate variables, the average nominal price paid to milk producers, the average wholesale price of dairy products and monthly milk production.

Coefficients	Dependent Variable		
	PPPL	PMDA	PRODUCTION
PPTM_1	0,0792993	0,150467	0,117575
	0,00069***	0,01779**	<0,00001***
TEMP_1	2,73405	9,7785	4,13553
	0,06158*	0,01575**	0,00014***
PPPL_1	-	-0,286589	0,0350027
	-	0,01585**	0,25691
PMDA	0,359476		-0,000654275
	<0,00001***		0,97861
PRODUCTION_1	0,10304	0,1631	-
	0,08464*	0,00460***	-
S1_1	-35,0392	-169,223	-4,6638
	0,12885	0,00841***	0,77876
S2_1	-56,6594	-201,664	-31,2479
	0,01547**	0,00192***	0,06335*
S3_1	-61,4147	-222,969	-61,9451
	0,01992**	0,00235***	0,00132***

*** Significant at 1%, ** at 5% and * at 10%.

Source: Adapted from GRETL, 2015.

The dependent variable nominal average price paid to milk producers was related to most of the significant variables. However, the relationships that have shown the greatest relevance to the theme proposed in the paper are those with the climate series and seasonality.

The relationship with climatic variables shows that, at least empirically, an increase in total monthly rainfall and average monthly temperature has a positive impact on the average nominal price paid to milk producers in a one-month lag.

The most significant impact of this analysis is in relation to temperature, in which it was possible to verify that a 1% increase in temperature can have a positive impact of approximately 2.73% on the average nominal price paid to dairy farmers, demonstrating that increases in temperature within the climatic range analysed are beneficial to the price paid to dairy farmers.

In a shock-response model, as shown in Graph 6, it can be considered that the impacts of an increase in temperature and rainfall on the price paid to producers are positive, i.e. they can have a decisive impact on the income of dairy farmers for up to 12 months.

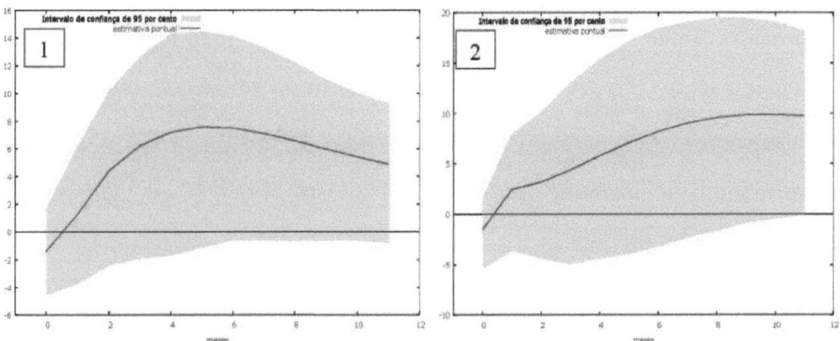

Graph 6 - Response of the Nominal Price Paid to Milk Producers to a shock in temperature (1) and rainfall (2).

Source: Prepared by the authors, 2015.

The response to the temperature shock is most pronounced between the third and fifth month, after which the effects of the shock diminish.

The effects on the price paid to dairy farmers in response to the shock in rainfall are particularly evident from the fifth to the ninth month, after which the effects of the shock stabilise.

The relationship between the average nominal price paid to dairy farmers and seasonality shows that the period between March and October is when the price can absorb the impact of this phenomenon. At least empirically, seasonality can have a negative impact of more than 50 per cent on the average nominal price paid to milk producers.

These factors are extremely important for understanding which variables have an impact on dairy farmers' decisions to remain in dairy farming and how they influence the technological model adopted by producers in this sector.

Among the dependent variables analysed was the average wholesale price of dairy products - whole raw milk, which showed significant relationships with all the variables. However, the climate series and seasonal dummies were the most significant among the relationships.

Empirically, Table 5 shows that a 1 per cent increase in total monthly rainfall has a positive impact of 0.15 per cent on the average price of whole raw milk.

For temperature, the analysis shows that, at least empirically, a 1% increase in the average monthly temperature has a positive impact of 9.77% on the average wholesale price of dairy products.

A shock-response model, as shown in Graph 7, shows that the impacts on the average wholesale price of dairy products of an increase in temperature and rainfall are positive.

The response to the shock can be deterministic in the sector's activities, such as the decision on the level of investment by wholesale industries in the sector, in new technologies, expanding the production of products and adapting the production model to new environmental and economic trends, within up to 12 months.

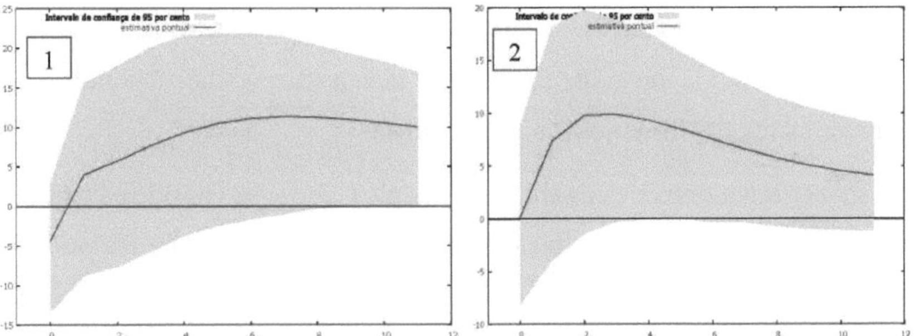

Graph 7 - Response of the Average Wholesale Price of Milk Products to a Shock in Precipitation (1) and Temperature (2).

Source: Prepared by the authors, 2015.

The average wholesale price of dairy products in response to the rainfall shock is highlighted from the fourth to the sixth month and after this period the effects of the shock stabilise.

The response to the shock in the sense of temperature is evident from the first to the fourth month, after which the effects of the shock are reduced.

Climate change, i.e. an increase or decrease in rainfall and/or temperature, at least historically, has had a positive influence on the average wholesale price of dairy products. Especially in the case of whole milk, this impact is relatively important given the fact that this product makes up the basic food basket of most Brazilian families.

The impact of seasonality on the wholesale price of milk is felt in every month of the year. The effects of this shock are not positive for this dependent variable, and can have a negative impact of up to 222% in the months of May to August.

This shows the importance of climatic variables and seasonality in establishing a pattern, stabilisation or even an increase in the price paid to dairy farmers and the price paid by industries for litres of milk used in the production of dairy products.

Another point to note is the intrinsic relationship between milk production and the climate series shown in Table 5. It is considered that, at least empirically, a 1% increase in rainfall has a positive impact of approximately 0.12% on milk production in a month.

The response to a 1% increase in temperature, empirically speaking, can have a positive impact of 4.13% on milk production, demonstrating the importance of temperature for increasing the production of this input.

In a shock-response model, as shown in Graph 8, it can be considered that the impacts on milk production from an increase in rainfall and temperature are positive and can affect the volume of milk produced and, consequently, its derivatives for up to 12 months. The response to the shock is more significant in terms of precipitation in the first month, after which the effects of the shock are reduced until they become constant.

As for the response of milk production to the temperature shock, the effects can interfere with the production of the input for up to 2 months afterwards. After this response, production is no longer affected by the shock and becomes stable.

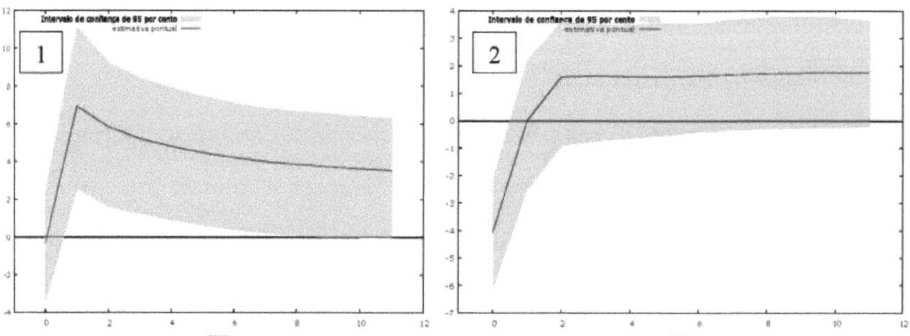

Graph 8 - Response of Milk Production to a Shock in Precipitation and Temperature.

Source: Prepared by the authors, 2015.

It is well known that high temperatures are directly related to a decrease in production, due to the negative effects they have on the gestation and productive efficiency of dairy cows and the heat stress caused to cattle exposed to these temperatures.

It is therefore of the utmost importance to analyse the results in detail, since, as mentioned above, they were determined using the temperatures provided in the research data during the period of time studied, which ranged from 18°C to 26°C.

In this reasoning, the results imply, at least empirically and historically, that increasing the temperature from 18°C to 19°C or from 25°C to 26°C is beneficial for milk production.

Milk production showed a significant relationship with seasonal dummies 3, showing that the impact of seasonality on milk production is felt negatively by approximately 62% in the months of May to August.

This means that if climatic variations, at least historically, influence the formation of the producer's income

The regular rainfall and temperatures considered optimal for dairy farming are essential for the smooth running of the milk production chain.

If the Brazilian climate scenario changes, such as longer periods of drought and high temperatures, and thus presents a broad base of transmission, i.e. increases capacity in subsequent months, it could pose a risk to the continuity of milk production and the quality of the products derived from it. In addition, this scenario is likely to contribute to a worsening food security situation in the country, considering the importance of milk as a source of nutrients necessary for a healthy life.

This debate can be considered the main target of this research, due to the imminent concern about climate change, largely caused by anthropogenic activities and its consequences in the area of food production and the formation and constitution of public stocks. This is not just a purely economic issue, due to losses in livestock and production and the slowdown in the industrial sector. It can directly affect the way in which public policies should be conducted to guide strategies for maintaining milk stocks in order to guarantee food security, as well as helping producers to cope with variations in the climate.

CHAPTER 4

FINAL CONSIDERATIONS

The transmission capacity of the effects of the climate series was found, with greater evidence for the historical series of average monthly rainfall and average monthly temperature. It was therefore possible to discuss the transmission capacity of the impacts on the average nominal price paid to milk producers, on the average wholesale price of dairy products - whole raw milk and on milk production, based on the results of the VAR (Vector Auto Regression) model.

These impacts are of paramount importance in order to assess the dependence of the production chain on the effects of climate change and the extent to which this can have an impact on guaranteeing food security.

However, more in-depth research is needed to measure this dependence in the economic, social and environmental spheres, considering other variables, regions and periods, for example, focusing on the consequences for the productive sector and the impacts on the continuity of policies aimed at food security.

CHAPTER 5

REFERENCES

BRAZILIAN ACTION FOR NUTRITION AND HUMAN RIGHTS. **Human right to adequate food in the context of food and nutrition security**. Brasília, 2010. Available at : < http://www.actuar-acd.org/uploads/5/6/8/7/5687387/dhaa_no_contexto_da_san.pdf> Accessed 21 Apr. 2015

ALENCAR, Álvaro Gurgel. From the strategic concept of food security to the FAO action plan to combat hunger. **REVISTA BRASILEIRA DE POLÍTICA INTERNACIONAL**. v.44, n.1, jan./jun. 2001.

ALMEIDA, Fabrício Pelizer. **Dairy farming: the state of dairy farming in Brazil and worldwide**. Agribusiness Management Technology. Uberaba, 2007. v.2, p. 27-50.

BACCARI JÚNIOR, FLAVIO. Environmental management for milk production in hot climates. In: **CONGRESSO BRASILEIRO DE BIOMEDICINA**, 2., 1998, Goiânia : Universidade Católica de Goiás, 1998. p. 136 - 161.

METEOROLOGICAL DATABASE FOR TEACHING AND RESEARCH. Available at : <http://www.inmet.gov.br/projetos/rede/pesquisa/> Accessed on : 06 Jun. 2015.

BRAZIL. Law no. 11.346, of 15 September 2006. Creates the Food and Nutrition System - SISAN with a view to ensuring the human right to adequate food and makes other provisions. **Federal Official Gazette**, 18 September 2006. Available at: <http://www.planalto.gov.br/ccivil_03/_ato2004- 2006/2006/lei/l11346.htm> Accessed on: 25 Feb.2015.

BRAZIL. Ministry of the Environment. Secretariat for Climate Change and Environmental Quality. **MMA technical panel report** on financing, benefits and co-benefits. Jun. 2012. Available at: <file:///C:/Users/Lara/Downloads/redd+_relat%C3%B3rio_de_painel_t%C3%A9cnico_do_mma.pdf> Accessed on: 12 Apr. 2015.

BRAZIL. Decree No. 30.691/1952, of 29 March 1952. Approves the new Regulations for the industrial and sanitary inspection of products of animal origin - RISPOA. **Federal Official Gazette**. Available at: < http://www.agricultura.gov.br/arq_editor/file/Aniamal/MercadoInterno/Requisitos/RegulamentoInspecaoIndustrial.pdf> Accessed on: 14 Apr. 2015.

CENTRE FOR ADVANCED STUDIES IN APPLIED ECONOMICS. Available at:

<http://cepea.esalq.usp.br/leite/> Access: 06 Jun. 2015.

MILK INTELLIGENCE CENTRE. **About CILEITE**. Available at: <http://www.cileite.com.br/node/14> Accessed on: 01 October 2015.

NATIONAL SUPPLY COMPANY. Available at: <http://www.conab.gov.br/conteudos.php?a=1252&t=> Accessed on: 06 Jun. 2015.

DICKEY, D. A. & FULLER, W. A. (1979). **Distribution of the estimator for auto-regressive time series with a unit root.** Journal of the *American Statistical Association,* 74(366):427-431.

BRAZILIAN INSTITUTE OF STATISTICAL GEOGRAPHY. **Livestock Production Statistics**. Sept. 2014. 50 p. Available at: <http://ibge.gov.br/home/estatistica/indicadores/agropecuaria/producaoa gropecuaria/abate-leite-couro-ovos_201402_publ_completa.pdf> Accessed on 29 Mar. 2015.

BRAZILIAN INSTITUTE OF GEOGRAPHY AND STATISTICS. **Municipal Livestock 2013**. Rio de Janeiro: IBGE, vol. 41, p.1-108, 2013.

Institute for Applied Economic Research. **First social analysis of the new National Household Sample Survey, 2013**. Available at <http://www.ipeadata.gov.br/> Accessed on: 14 Oct. 2015.

AMAZON RESEARCH INSTITUTE. **What is Brazil's contribution to climate change? And what is the profile of Brazilian emissions?** Available at <http://www.ipam.org.br/saiba- mais/abc/mudancaspergunta/Qual-a-contribuiçãoo-do-Brasil-para-as- mudancas-climaticas-E-qual-o-perfil-das-emissoes-brasileiras-/27/17> Accessed on: 22 Apr. 2015.

INTERGOVERNMENTAL PANEL ON CLIMATE CHANCE. Available at: <www.ipcc.ch> Accessed on: 26 Feb. 2015.

BRAZILIAN AGRICULTURAL RESEARCH COMPANY. **Bioclimatology applied to dairy cattle production in the tropics**. Teresina, Mar. 2009. 83p.

BRAZILIAN AGRICULTURAL RESEARCH COMPANY. Dairy farming in the Bragantina Zone. **Milk production chain.** Dec, 2005. Available at: http://sistemasdeproducao.cnptia.embrapa.br/FontesHTML/Leite/GadoL eiteiroZonaBragantina/paginas/cadeia.htm > Accessed on: 01 October 2015.

ENDERS, W. **Applied econometric time series**. New York: John Wiley and Sons, Inc., 1995.

FIGUEIREDA, Sérgio Rangel; BELIK, Walter. Transformations in the industrial link of the milk production chain. **Cadernos de Debate magazine**, v. 7, p. 31-34, 1999.

FOLEY, Jonathan A., et al. **Solutions for a cultivated planet**. Vol. 478, p. 337 - 342, Oct. 2011.

GUJARATI, Damodar N. **Basic Econometrics**. New York: McGraw-Hill, 2000. 846 p.

GRETL - GNU REGRESSION, ECONOMETRICSAND TIME-SERIES LIBRARY, 1.10.1, 2015).

HAHN, G. Leroy. **Bioclimatology and zootechnical facilities**: theoretical and applied aspects. Jaboticabal: FUNEP, 1993. 28 p.

MENDOÇA, Francisco; DANNI-OLIVEIRA, Inês Moresco. **Climatology: basic notions and climates of Brazil**. São Paulo: Oficinas de Textos, 2007. 206 p.

MACEDO, Dione Chaves; TEIXEIRA, Estelamar Maria Borges; JERÔNIMO, Marlene; BARBOSA, Ozeni Amorim; OLIVEIRA, Maria Rita Marques. The construction of food and nutrition security policy in Brazil. **Revista Simbio-Logias**, V.2, n.1, May/2009.

MOREIRA, Igor. Geographical Space - **General Geography of Brazil**. São Paulo, SP: Ática, 2002.

OMETTO, José Carlos. **Plant bioclimatology**. São Paulo: Ed. Agronômica Ceres, 1981. 425 p.

PORTAL BRASIL. Economy and Employment. **Brazil launches plan to double milk exports in 3 years**. Available at: <http://www.brasil.gov.br/economia-e-emprego/2015/09/brasil-lanca- plano-para-dobrar-exportacao-de-leite-em-3-anos> Accessed on: 02 Oct. 2015.

PIRES, Maria de Fátima Ávila; VILELA, Duarte; VERNEQUE, Rui da Silva; TEODORO, Roberto Luiz. **Reflexes of heat stress on the behaviour of lactating cows**. In: BRAZILIAN SYMPOSIUM ON ENVIRONMENT IN MILK PRODUCTION, 1., 1998, Piracicaba.
Milk production in hot climates. Piracicaba: FEALQ, 1998. P. 68 - 102.

REIS, José. **Scientists discuss the return of pests**. Folha de São **Paulo, São Paulo, 1 December 1996. Periscope" column, p. 5-6.**

UNION OF THE DAIRY AND DAIRY PRODUCTS INDUSTRY OF PARANÁ. **Benefits of milk and its derivatives**, 2013. Available at: <http://www.fiepr.org.br/sindicatos/sindileitepr/beneficios-do-leite-e-seus-derivatives13405197525.shtml> Accessed: 29 Sep. 2015.

SILVA, Thieres George Freire, et al. Impacts of climate change on dairy production in the state of Pernambuco: analysis for the B2 and A2 scenarios of the IPCC. **Revista Brasileira de Meteorologia**, v.24, n.4, p 489 - 501, dec. 2009.

ALMEIDA, Sandra. Live Healthy. **The importance of milk and its derivatives in the diet, 2011**. Available at <http://vivasaudavel2010.blogspot.com.br/2011/04/importancia-do-leite-eseusderivados.html> Accessed on: 29 Sep. 2015.

VALENTE, Flávio Luiz Schieck. **From the fight against hunger to food and nutritional security: the right to adequate food**. R. Nutro. PUCCAMP, Campinas. jan./jun. 1997. Available at: < http://portal.revistas.bvs.br/index.php?search=Rev.%20nutr.%20PUCCA MP&connector=ET&lang=en> Accessed 24 Mar. 2015.

VIANELLO, Rubens Leite; ALVES, Adail Rainier. **Basic meteorology and applications**. 2. ed. Viçosa: Ed. UFV, 2012. 460 p.

CHAPTER 6

APPENDIX
Database: Historical Series

Period	PPTM	TEMP	PPPL	PMDA	PRODUCTION	S1	S2	S3
Jan/05	247,7	24,7	534,57	550,0	529,4	0	1	1
Feb/05	124,8	24,7	543,25	560,0	535,2	0	1	1
Mar/05	282,7	24,5	552,29	580,0	553,3	1	0	1
Apr/05	34,3	24,2	572,07	640,0	576,1	1	0	1
May/05	31,0	21,3	587,19	670,0	593,4	1	1	0
Jun/05	0,7	20,1	590,37	630,0	605,0	1	1	0
Jul/05	0,1	19,4	582,66	570,0	620,6	1	1	0
Aug/05	22,2	23,0	530,76	470,0	609,6	1	1	0
Sep/05	27,0	23,5	500,72	490,0	595,1	1	0	1
Oct/05	46,1	26,2	484,41	450,0	577,4	1	0	1
Nov/05	301,0	23,5	468,84	460,0	554,7	0	1	1
Dec/05	317,3	23,6	447,72	450,0	541,6	0	1	1
Jan/06	171,5	24,8	424,69	460,0	543,6	0	1	1
Feb/06	142,2	25,1	444,29	510,0	549,5	0	1	1
mar/06	276,9	24,2	453,55	530,0	568,2	1	0	1
Apr/06	75,4	23,0	483,98	570,0	591,5	1	0	1
May/06	12,1	19,6	497,24	560,0	609,4	1	1	0
Jun/06	0,1	18,9	501,97	540,0	621,3	1	1	0
Jul/06	0,9	19,3	504,02	530,0	637,3	1	1	0
Aug/06	9,5	22,0	502,76	530,0	626,0	1	1	0
Sep 06	53,0	22,7	512,81	521,0	611,0	1	0	1
Oct/06	200,0	23,5	513,48	529,4	592,9	1	0	1
Nov/06	243,6	23,6	510,22	542,5	569,6	0	1	1
Dec/06	318,7	24,2	506,30	550,0	556,1	0	1	1
Jan/07	381,7	24,3	503,60	539,4	557,4	0	1	1
Feb/07	221,9	24,5	521,80	560,0	563,5	0	1	1
mar/07	58,0	25,1	541,64	584,4	582,7	1	0	1
Apr/07	53,5	23,7	560,07	640,6	606,6	1	0	1
May/07	3,7	20,5	593,98	714,6	624,9	1	1	0
Jun/07	0,4	19,6	630,98	734,9	637,1	1	1	0
Jul/07	6,7	19,8	690,29	860,2	653,5	1	1	0
Aug/07	0,1	21,1	786,32	923,8	641,9	1	1	0
Sep 07	5,1	24,4	826,67	880,9	626,6	1	0	1
Oct/07	50,4	25,9	814,43	806,1	608,0	1	0	1
Nov/07	134,4	24,9	745,95	773,2	584,1	0	1	1
Dec/07	197,5	24,9	721,51	731,4	570,3	0	1	1
Jan/08	297,7	23,9	706,18	740,0	586,7	0	1	1
Feb/08	191,1	24,3	722,10	769,9	593,2	0	1	1
Mar 08	208,8	23,8	747,48	775,4	613,3	1	0	1
Apr/08	110,2	23,7	761,42	842,5	638,5	1	0	1

May/08	11,9	20,2	777,66	833,9	657,7	1	1	0
Jun/08	0,1	20,3	779,38	801,8	670,6	1	1	0
Jul/08	0,1	18,9	758,88	744,6	687,8	1	1	0
Aug/08	4,3	22,2	732,15	716,6	675,6	1	1	0
Sep 08	54,9	23,2	664,21	681,5	659,6	1	0	1
Oct 08	35,0	25,8	631,35	608,0	640,0	1	0	1
Nov/08	227,7	24,4	591,70	598,7	614,8	0	1	1
Dec 08	445,6	23,8	593,11	574,0	600,3	0	1	1
Jan 09	223,2	24,5	588,67	611,3	607,7	0	1	1
Feb 09	197,4	25,0	591,35	631,7	614,4	0	1	1
Mar 09	197,1	24,8	605,74	650,9	635,2	1	0	1
Apr 09	79,2	22,8	633,84	671,1	661,3	1	0	1
May 09	33,4	21,2	667,21	778,4	681,3	1	1	0
Jun/09	24,0	18,9	698,02	865,4	694,6	1	1	0
Jul 09	0,2	20,8	765,23	871,9	712,4	1	1	0
Aug/09	24,9	21,3	770,48	837,8	699,8	1	1	0
Sep 09	61,4	24,3	750,69	755,7	683,1	1	0	1
Oct 09	186,0	24,4	708,82	659,0	662,9	1	0	1
nov/09	143,2	25,6	649,70	629,2	636,8	0	1	1
Dec 09	309,0	23,9	617,60	633,9	621,7	0	1	1
Jan/10	194,9	25,0	609,90	650,0	642,7	0	1	1
Feb/10	97,9	25,7	643,10	776,6	649,8	0	1	1
Mar/10	224,9	24,6	700,20	869,8	671,8	1	0	1
Apr/10	82,7	22,4	784,00	913,4	699,4	1	0	1
May/10	22,4	19,8	813,80	828,6	720,5	1	1	0
Jun/10	11,3	18,2	812,20	751,4	734,6	1	1	0
Jul/10	0,1	20,4	755,00	764,4	753,5	1	1	0
Aug/10	0,1	20,6	705,10	730,9	740,1	1	1	0
Sep/10	36,2	23,8	718,10	739,8	722,5	1	0	1
Oct/10	164,2	24,0	716,40	811,7	701,1	1	0	1
Nov/10	243,0	23,5	725,60	814,1	673,5	0	1	1
Dec/10	273,5	24,9	723,50	808,0	657,6	0	1	1
Jan/11	236,0	24,2	721,10	791,4	670,9	0	1	1
Feb/11	71,6	24,3	728,50	801,0	678,3	0	1	1
Mar/11	466,5	23,0	760,40	830,5	701,3	1	0	1
Apr/11	65,9	22,8	800,20	875,5	730,1	1	0	1
May/11	0,5	20,0	837,90	915,5	752,1	1	1	0
Jun/11	12,6	18,3	864,60	908,4	766,8	1	1	0
Jul/11	0,1	19,7	871,18	942,9	786,6	1	1	0
Aug/11	0,1	21,9	882,70	974,7	772,6	1	1	0
Sep/11	1,0	24,1	906,50	956,8	754,2	1	0	1
Oct/11	135,6	23,3	896,20	905,1	731,8	1	0	1
Nov/11	182,3	23,1	860,60	865,6	703,0	0	1	1
Dec/11	328,8	23,5	840,00	839,1	686,4	0	1	1
Jan/12	246,1	23,4	822,00	860,0	682,4	0	1	1

Feb/12	156,0	24,8	838,80	863,6	689,9	0	1	1
Mar/12	128,5	24,2	864,00	865,2	713,3	1	0	1
Apr/12	52,4	23,4	882,80	898,7	742,6	1	0	1
May/12	44,0	21,4	891,60	873,9	765,0	1	1	0
Jun/12	26,7	22,1	867,90	883,4	779,9	1	1	0
Jul/12	0,2	20,7	860,60	897,7	800,0	1	1	0
Aug/12	0,1	22,2	871,90	902,3	785,8	1	1	0
Sep/12	16,7	23,8	889,55	923,0	767,1	1	0	1
Oct/12	32,5	26,2	896,00	988,9	744,4	1	0	1
Nov/12	271,4	24,6	908,88	1020,0	715,1	0	1	1
Dec/12	157,9	25,9	899,23	960,0	698,2	0	1	1
Jan/13	271,6	24,4	892,54	929,0	715,2	0	1	1
Feb/13	115,9	24,9	904,97	975,6	736,4	0	1	1
Mar/13	193,3	24,4	930,08	1040,0	757,3	1	0	1
Apr/13	106,7	22,1	971,60	1057,1	765,6	1	0	1
May/13	37,7	20,9	1006,92	1144,0	808,9	1	1	0
Jun/13	21,9	20,8	1038,48	1231,9	805,9	1	1	0
Jul/13	0,1	19,8	1071,39	1313,9	835,6	1	1	0
Aug/13	0,6	20,9	1116,83	1277,0	818,8	1	1	0
Sep/13	68,5	23,2	1155,01	1312,9	799,4	1	0	1
Oct/13	134,8	23,8	1157,33	1264,9	772,4	1	0	1
Nov/13	160,3	24,4	1134,05	1089,4	746,0	0	1	1
Dec/13	268,2	24,4	1063,46	954,4	739,9	0	1	1
Jan/14	114,1	25,3	996,69	927,3	730,9	0	1	1
Feb/14	96,8	25,3	1008,47	1018,6	755,4	0	1	1
Mar/14	114,1	24,6	1056,98	1250,8	785,6	1	0	1
Apr/14	100,8	23,1	1117,31	1282,6	799,4	1	0	1
May/14	2,2	20,0	1123,91	1186,9	848,0	1	1	0
Jun/14	0,3	19,8	1105,27	1198,9	819,9	1	1	0
Jul/14	39,6	19,0	1113,06	1205,9	866,2	1	1	0
Aug/14	0,7	21,4	1119,05	1206,4	838,0	1	1	0
Sep/14	7,6	24,3	1113,46	1211,9	804,7	1	0	1
Oct/14	67,7	25,6	1096,94	1119,8	793,1	1	0	1
Nov/14	208,3	24,0	1046,57	1025,0	758,4	0	1	1
Dec/14	256,8	24,1	993,16	898,1	741,6	0	1	1

I want morebooks!

Buy your books fast and straightforward online - at one of world's fastest growing online book stores! Environmentally sound due to Print-on-Demand technologies.

Buy your books online at
www.morebooks.shop

Kaufen Sie Ihre Bücher schnell und unkompliziert online – auf einer der am schnellsten wachsenden Buchhandelsplattformen weltweit! Dank Print-On-Demand umwelt- und ressourcenschonend produziert.

Bücher schneller online kaufen
www.morebooks.shop

Printed by Books on Demand GmbH, Norderstedt / Germany